How to use this book

A sample page

INSTRUCTION
What your child needs to do for the activity.

TITLE
The page title describes the skill your child will learn in these pages.

FUN ILLUSTRATIONS
Specially drawn illustrations which are fun, interesting and drawn at the right pedagogical level for your child.

COLOURFUL BORDERS
The page borders make each page as attractive as possible to stimulate your child.

EXAMPLE
The first one is done for you so you can show your child exactly what to do.

LOTS OF PRACTICE
Two pages where your child can practise and repeat the same skill to master it.

STICKERS
Place a sticker on each page as your child finishes.

WHO'S HIDING?
In each book, a little creature appears in the border of every double page so your child can have fun trying to find it.

EXTRA ACTIVITIES
Extra activities you might want to do with your child to further reinforce the skill or simply make it more enjoyable.

Step-by-step learning

STEP ONE — **Read** out the title of the activity page to your child.

STEP TWO — **Explain** the skill and show your child the example already done. **Make sure** they understand what to do. Your child will then have at least two pages to practise that same skill.

STEP THREE — **Help** your child put a **sticker** on the bottom of each page as they complete it.

Remember to be patient, encouraging and positive with your child, even when minor mistakes are made!

How to hold a pencil

It is important that you help your child hold his or her crayon or pencil in the correct way as shown here to ensure your child develops the right technique early on.

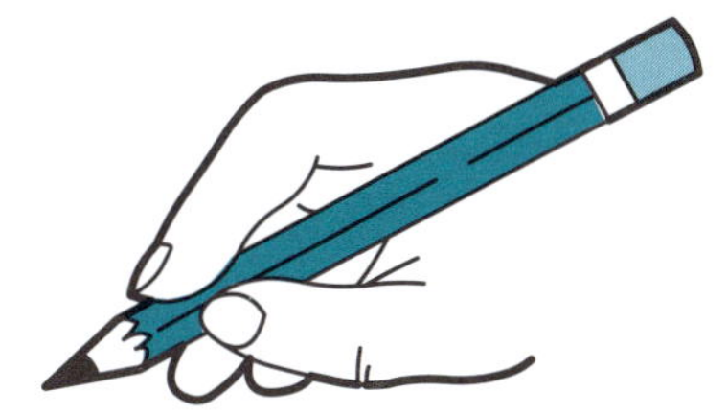

Counting and writing 1–10

Count the baby animals in each basket.
Write the number of animals in the box.

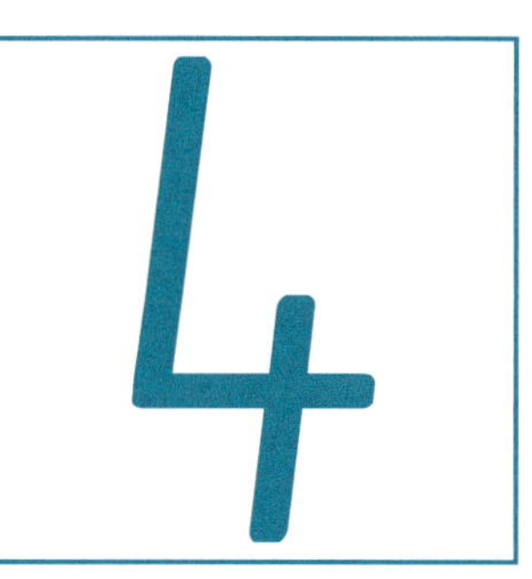

1. Before you start, go over counting to 10 with your child. Can they count to 10 on their fingers? To be able to add and subtract numbers, they will need to be confident in their counting.

Place a sticker here.

2. Ask your child to count all the rabbits on both pages. How many are there? How many puppies?

Place a sticker here.

Drawing 1-10 objects

Look at the number on each apple tree. Draw the matching number of apples.

1. Do you have a fruit bowl at home? Ask your child to count each different type of fruit in it.

Place a sticker here.

8

10

7

6

2. Ask your child to count the apples on one tree and collect the matching number of objects from the room.

Place a sticker here.

Adding 1 more

Draw 1 more biscuit in each jar. Count up how many biscuits there are now and write this number next to the jar.

4

1. Help your child to say a 'number story' for each jar — '3 biscuits and 1 more makes 4'.

Place a sticker here.

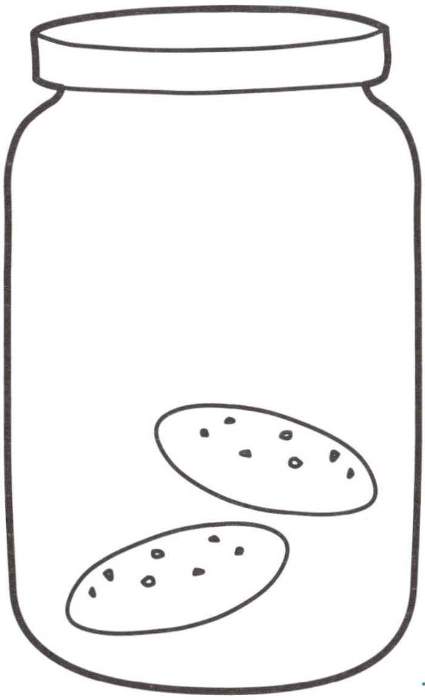

2. Ask your child which jar has the most biscuits now. Colour it in.

Place a sticker here.

Adding 2 more

How many people are in each group? Draw 2 more in each group, count how many there are altogether, then write the number on the card.

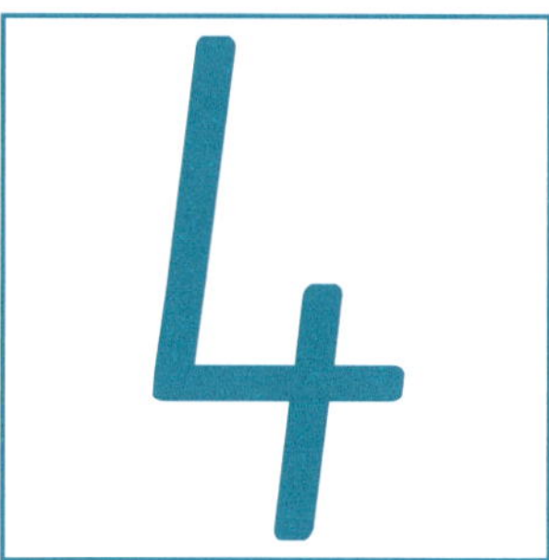

1. Ask your child to say a 'number story' for each row — '2 babies and 2 more makes 4'.

Place a sticker here.

2. The last row on this page has no pictures. Help your child to draw 2 faces, count them, and write the number in the box.

Place a sticker here.

Adding 3 more

How many pictures are in each group? Draw 3 more in each group, count how many there are altogether, then write the number on the card.

5

1. Ask your child to say a 'number story' for each row — '2 spoons and 3 more make 5'.

Place a sticker here.

2. The last row on this page has no pictures. Help your child to draw 3 plates, count them, and write the number in the box.

Place a sticker here.

Adding 4 more

How many pictures are in each group? Draw 4 more in each group, count how many there are altogether, then write the number on the card.

5

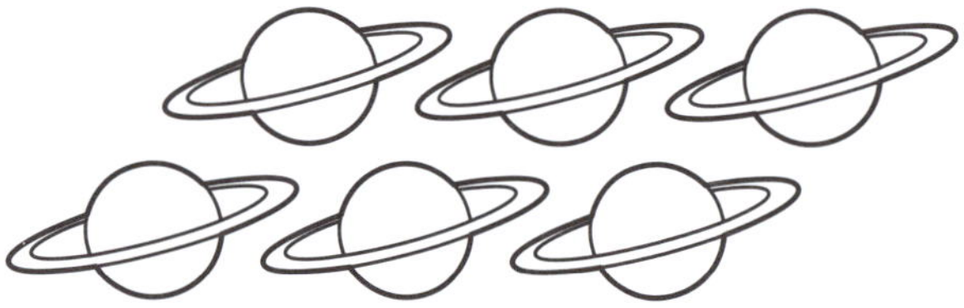

1. Don't forget to help your child make a 'number story' for each group. This becomes especially important as the numbers get larger.

Place a sticker here.

2. The last row on this page has no pictures. Help your child to draw 4 planets, then count them and write the number in the box.

Place a sticker here.

Adding 5 more

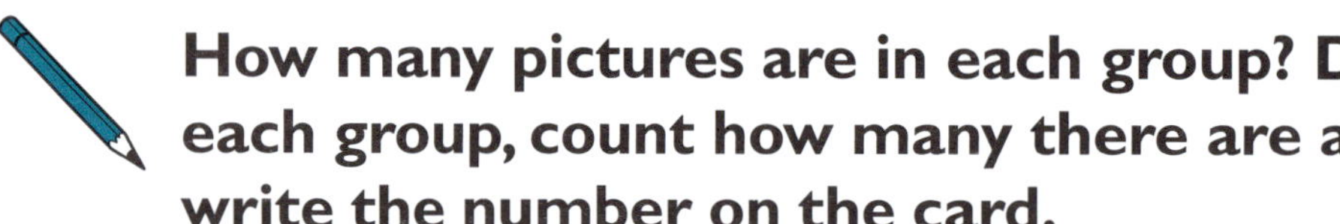

How many pictures are in each group? Draw 5 more in each group, count how many there are altogether, then write the number on the card.

7

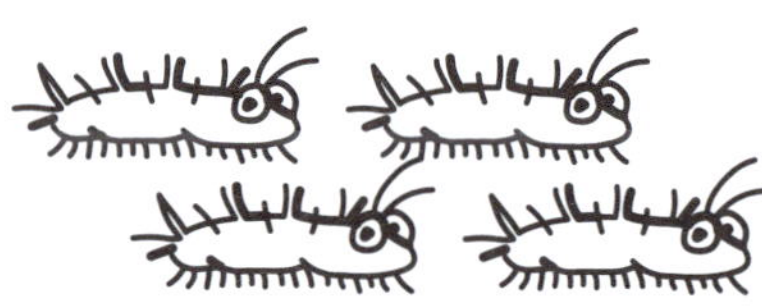

1. Ask your child to say a 'number story' for each row — '2 ants and 5 more makes 7'.

Place a sticker here.

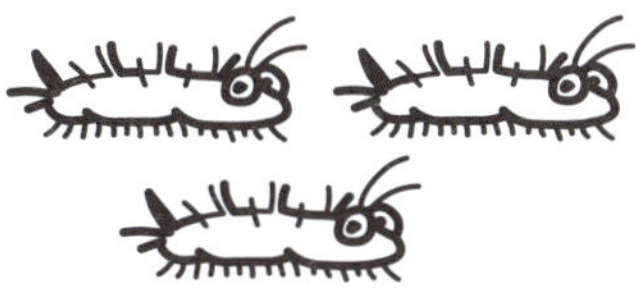

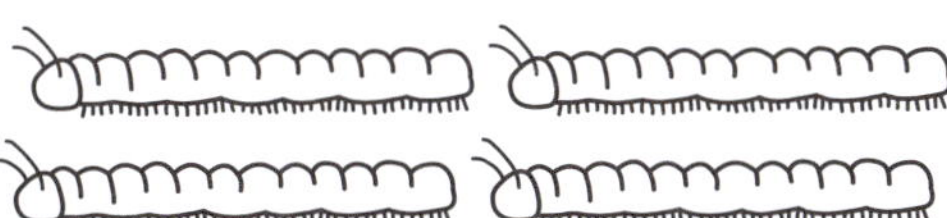

2. The last row on this page has no pictures. Help your child to draw 5 ants, count them, and write the number in the box.

Place a sticker here.

Taking 1 away

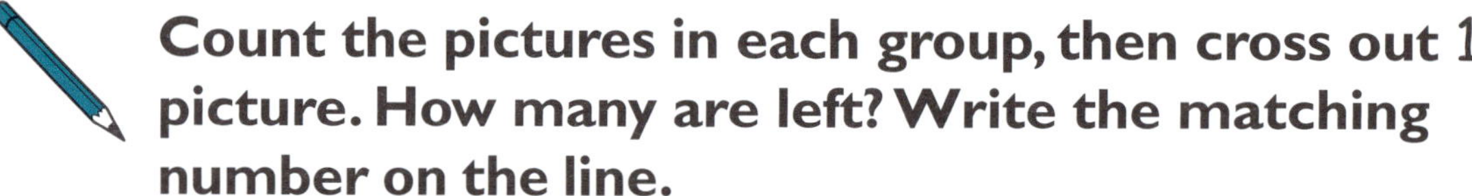
Count the pictures in each group, then cross out 1 picture. How many are left? Write the matching number on the line.

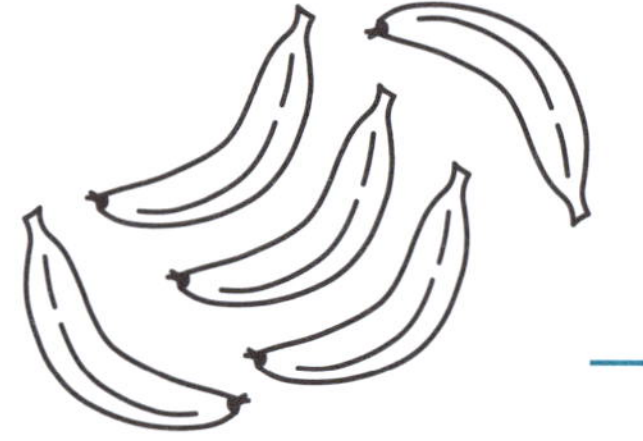

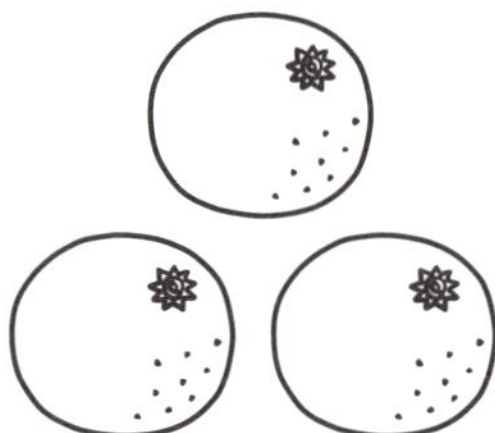

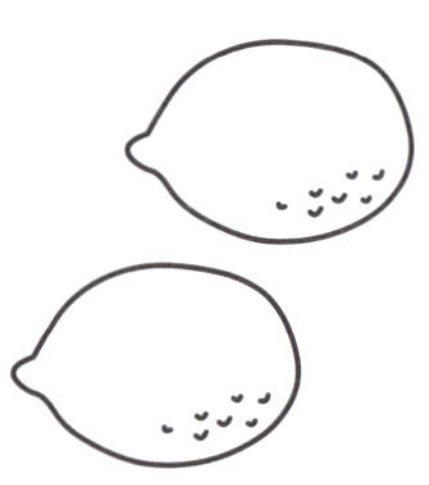

1. Help your child to say a 'number story' for each row — '4 apples take away 1 apple leaves 3 apples'.

Place a sticker here.

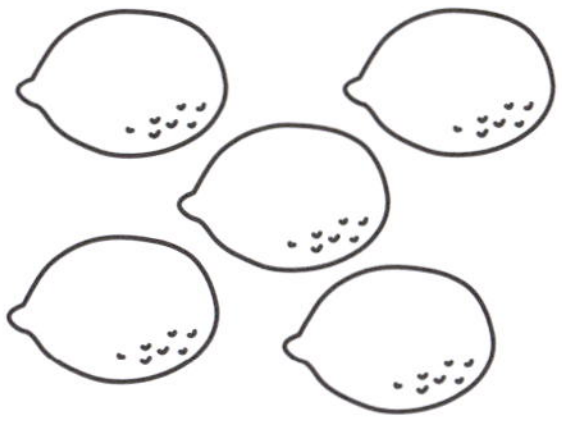

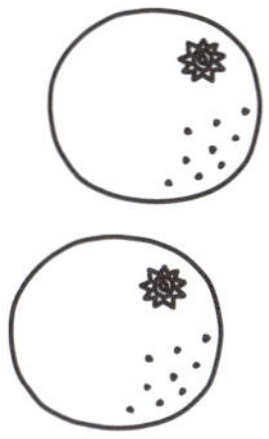

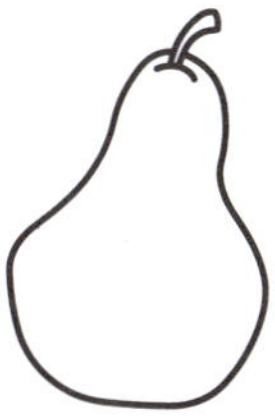

2. Is your child familiar with 'zero'? You may need to help them with the last example on this page where they need to write '0' on the line.

Place a sticker here.

Taking 2 away

Count the pictures in each group, then cross out 2 pictures. How many are left? Write the matching number on the line.

3 ______

______

______

______

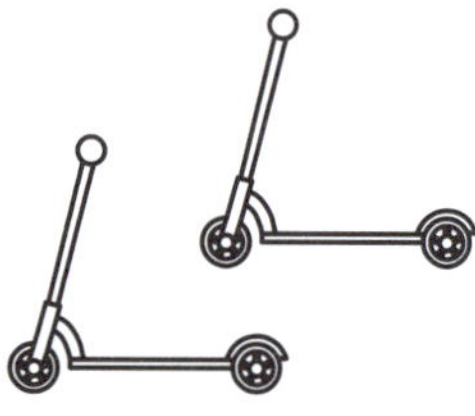

______

______

1. Help your child to say a 'number story' for each row — '5 cars take away 2 cars leaves 3 cars'.

Place a sticker here.

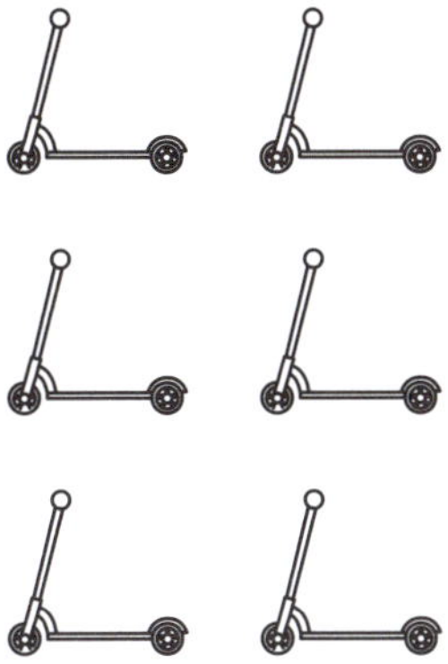

2. You can help your child understand 'taking away' by using real examples. For instance, if there are 4 bananas and you eat 2, how many are left?

Place a sticker here.

Taking 3 away

Count the pictures in each group, then cross out 3 pictures. How many are left? Write the matching number in the box.

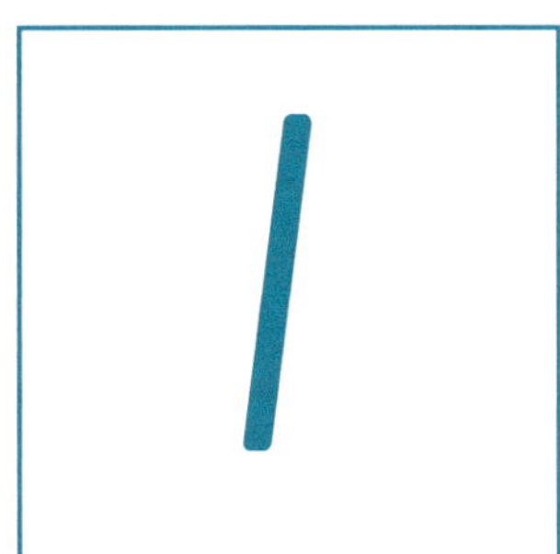

1. Ask your child to say a 'number story' for each row — '4 lions take away 3 lions leaves 1 lion'.

Place a sticker here.

2. You can do this activity using soft toys. Ask your child to count them, take 3 away, and count them again.

Place a sticker here.

Taking 4 away

Count the pictures in each group, then cross out 4 pictures.
How many are left? Write the matching number in the box.

1

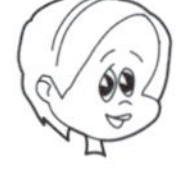

1. Ask your child to say a 'number story' for each row — '5 koalas take away 4 koalas leaves 1 koala'.

Place a sticker here.

2. Do you know any simple counting songs or rhymes you can sing with your child? See if you can find some that involve subtraction.

Place a sticker here.

Taking 5 away

Count the pictures in each group, then cross out 5 pictures. How many are left? Write the matching number in the box.

2

1. Ask your child to say a 'number story' for each row — '7 horses take away 5 horses leaves 2 horses'.

Place a sticker here.

2. Use a set of building blocks to do this activity. Ask your child to take a certain number away and count the remaining blocks.

Place a sticker here.

Adding 1 or taking 1 away

How many fish are there in each group? Write this number in the first card.
+ 1 means you have to draw 1 more fish, then count how many there are altogether. Write this number in the circle.

3 + 1 = 4

+ 1

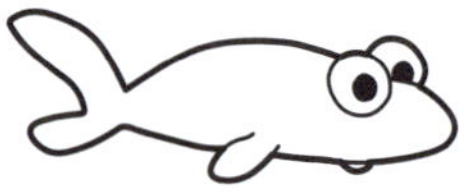

+ 1

1. Ask your child another adding question, for example, 'If I have 2 fish and I add 1 fish, how many fish do I have?' Your child can count it using their fingers.

Place a sticker here.

– 1 means you have to cross out 1 fish, then count how many fish are left. Write this number in the circle.

4 – 1 = 3

☐ – 1 = ◯

☐ – 1 = ◯

2. Try subtraction using a chalkboard or similar surface, eg. draw 5 pictures, rub out 1 and write how many are left.

Place a sticker here.

Adding 2 or taking 2 away

How many shells are there in each group? Write this number in the first card.
+ 2 means you have to draw 2 more shells, then count how many there are altogether. Write this number in the circle.

1	+	2	3
	+	2	
	+	2	

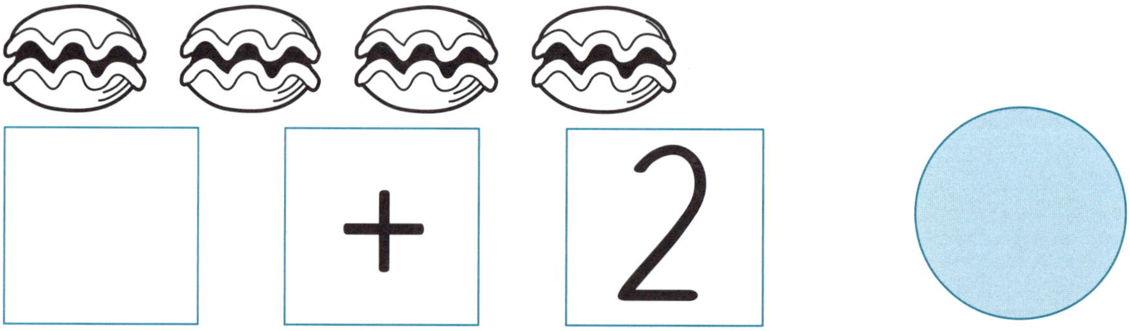

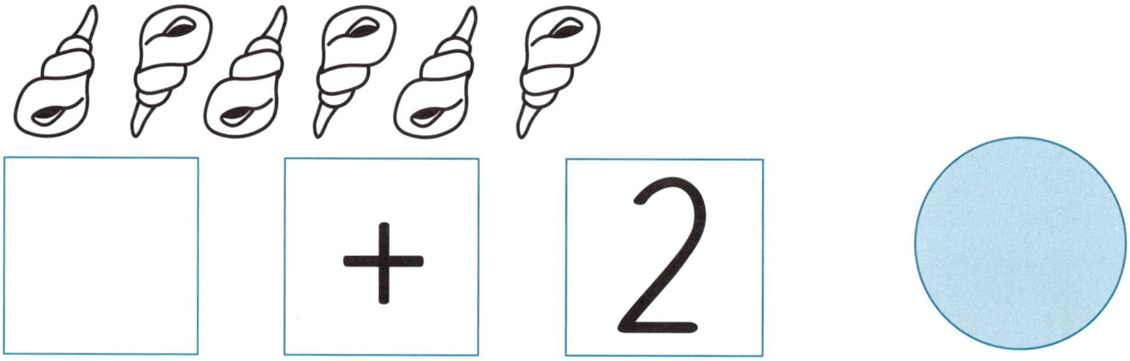

1. Ask your child another adding question, for example, 'If I have 3 shells and I add 2 shells, how many shells do I have?' Your child can count using their fingers.

Place a sticker here.

– 2 means you have to cross out 2 shells, then count how many shells are left. Write this number in the circle.

3 – 2 = 1

– 2

– 2

2. Use a set of playing cards (using the cards from 2 to 10) to help your child practise subtraction. Can they subtract one card from another and find the right number card for the anwer?

Place a sticker here.

Adding 3 or taking 3 away

How many vegetables are there in each group? Write this number in the first card.
+ 3 means you have to draw 3 more vegetables, then count how many there are altogether. Write this number in the circle.

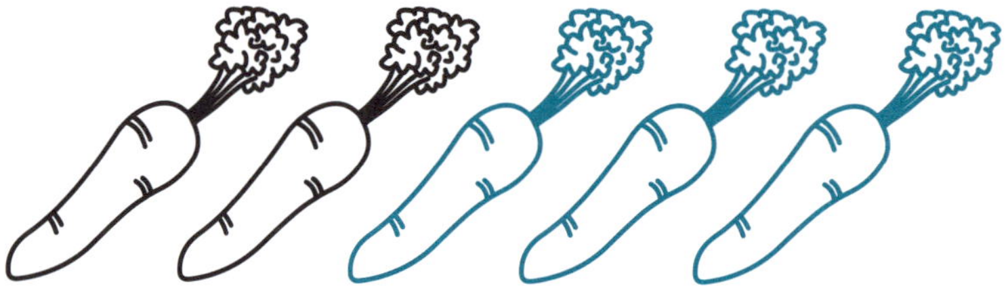

2	+	3	5
	+	3	
	+	3	

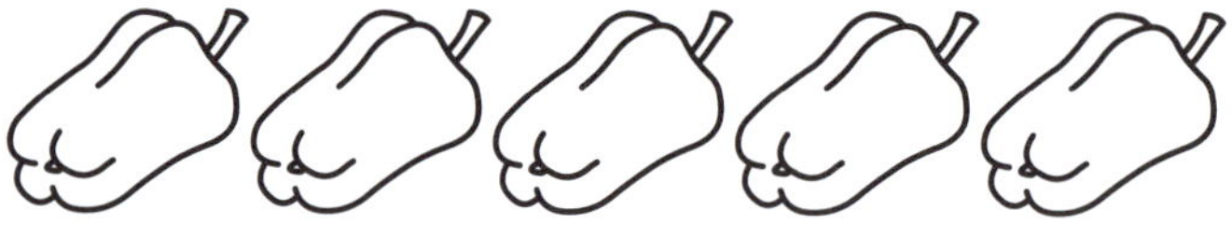

1. Ask your child another adding question, for example, 'If I have 5 peas and I add 3 peas, how many peas do I have?' Your child can count using their fingers.

Place a sticker here.

– 3 means you have to cross out 3 vegetables, then count how many vegetables are left. Write this number in the circle.

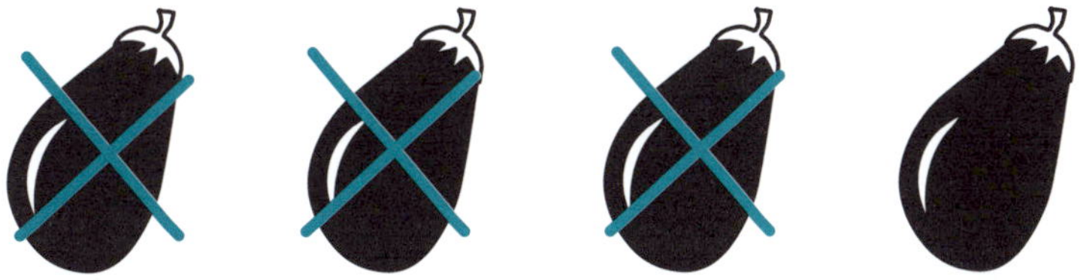

4 – 3 = 1

– 3

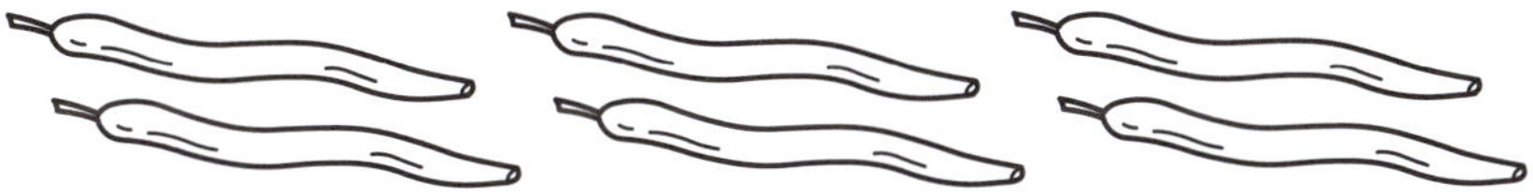

– 3

2. Use a set of building blocks or toys and have your child move them from one group to another and count them. If they take away 3 blocks, how many are left? If they add three, how many do they have?

Place a sticker here.

Adding 4 or taking 4 away

How many pictures are there in each group? Write this number in the first card.
+ 4 means you have to draw 4 more pictures, then count how many there are altogether. Write this number in the circle.

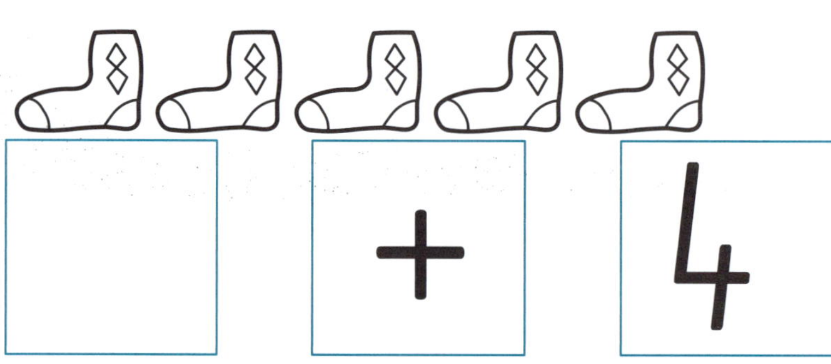

1. Ask your child another adding question, for example, 'If I have 4 shirts and I add 4 shirts, how many shirts do I have?' Your child can count it using their fingers.

Place a sticker here.

– 4 means you have to cross out 4 pictures, then count how many pictures are left. Write this number in the circle.

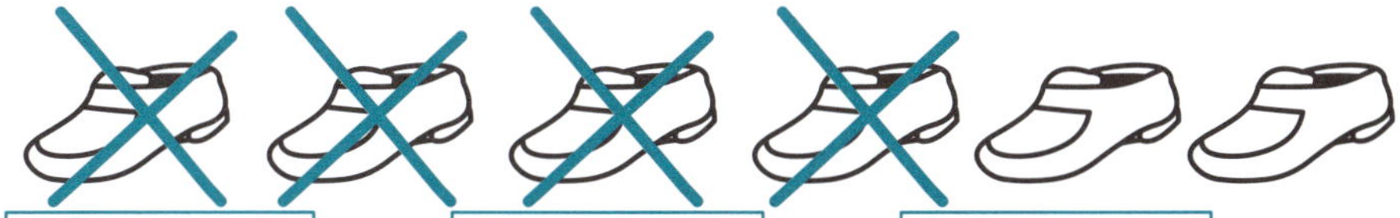

6 – 4 = 2

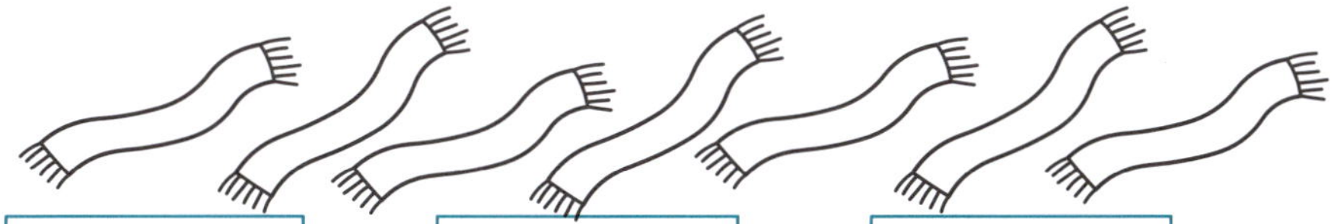

☐ – 4 = ◯

☐ – 4 = ◯

2. Use groups of everyday objects to practise adding and subtracting. The more your child can see how adding and subtracting are done, the easier it will be for them.

Place a sticker here.

Adding 5 or taking 5 away

How many aliens are there in each group? Write this number in the first card.
+ 5 means you have to draw 5 more aliens, then count how many there are altogether. Write this number in the circle.

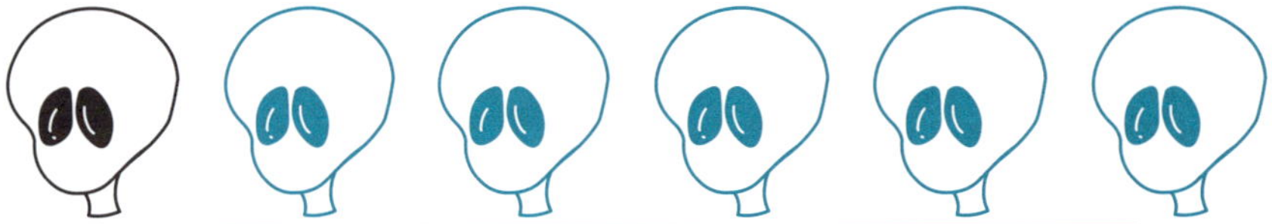

1	+	5	6

	+	5	

	+	5	

1. Ask your child another adding question, for example, 'If I see 4 aliens and along come 5 more aliens, how many aliens are there?' Your child can count it using their fingers.

Place a sticker here.

– 5 means you have to cross out 5 aliens, then count how many aliens are left. Write this number in the circle.

8	–	5	3
	–	5	
	–	5	

2. Ask your child another subtraction question, for example, 'If I see 10 aliens and 5 go into their spaceship, how many aliens are left?' Your child can count it using their fingers.

Place a sticker here.

Making groups of 5

Each pizza should have 5 olives on it. Count the olives on each pizza. If you need more olives, draw them in to make 5, then write your number story in the boxes. If there are too many olives, cross some out to leave 5 and write your number story in the boxes.

1	+	4
7	-	2

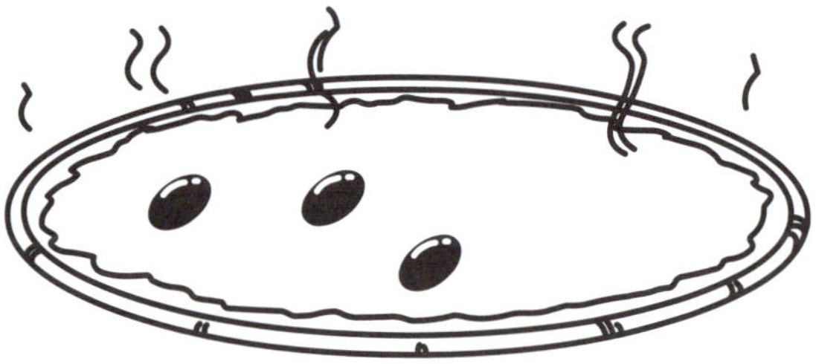

1. Help your child collect some counters (eg. buttons). Put them into 2 groups. Can they count how many in each group, then count how many altogether?

Place a sticker here.

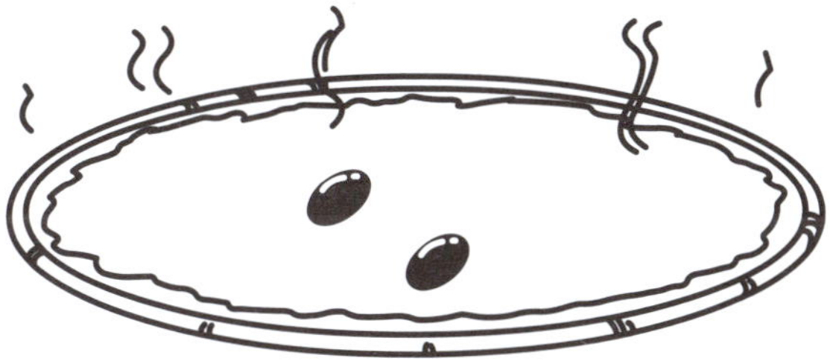

2. Help your child cut out pictures of food, then paste them into 2 groups. Discuss how many are in each group, and how many altogether.

Place a sticker here.

Making groups of 6

Each flower should have 6 petals on it. Count the petals on each flower. If you need more petals, draw them in to make 6, then write your number story in the boxes. If there are too many petals, cross some out to leave 6 and write your number story in the boxes.

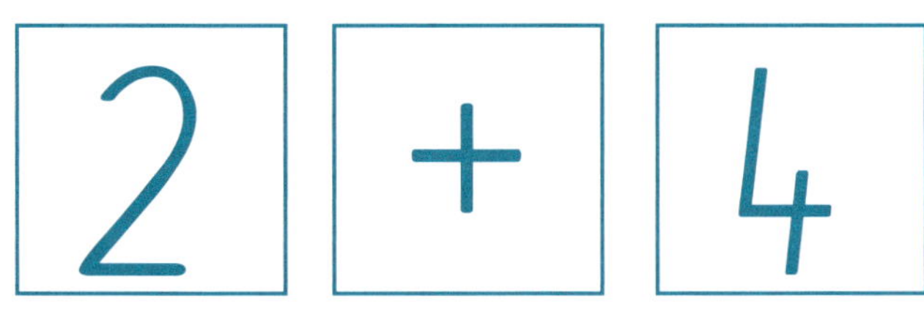

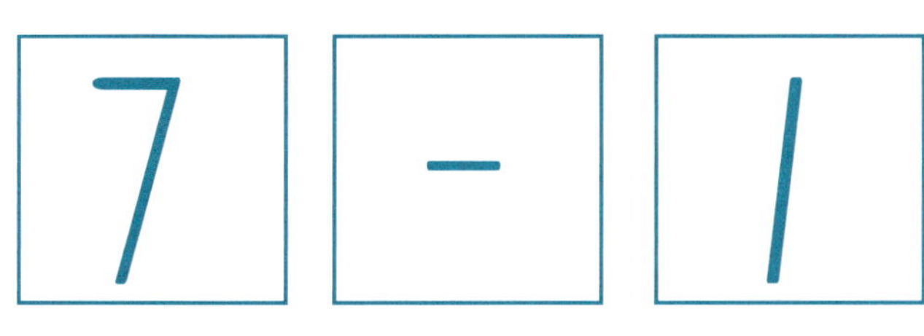

1. How else can you make groups of 6? Find objects in the garden and make groups of 6.

Place a sticker here.

2. Take a handful of leaves or other objects and count them. Take some away and ask your child to count how many are left.

Place a sticker here.

Making groups of 7

Here are some very spotty dogs! Each dog should have 7 spots on it. Count the spots on each dog. If you need more spots, draw them in to make 7, then write your number story in the boxes. If there are too many spots, cross some out to leave 7 and write your number story in the boxes.

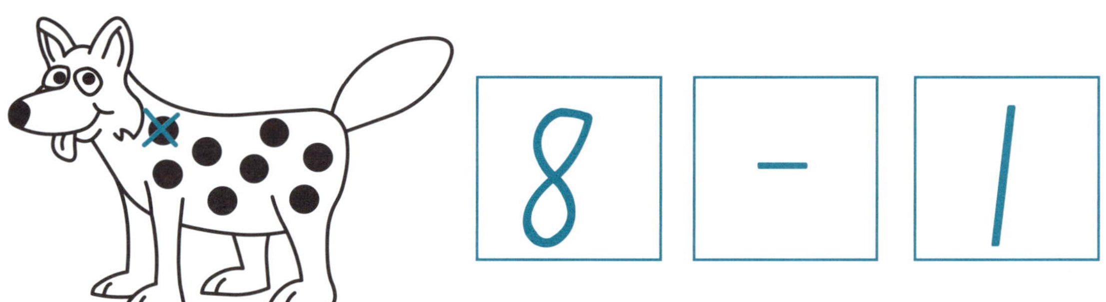

1. You could try this activity using a chalkboard and simple shapes. Your child can draw more shapes or rub some out to make a group of 7.

Place a sticker here.

2. Try to count the spots on a real cat or dog!

Place a sticker here.

Making groups of 8

How many legs does a spider need? Each spider should have 8 legs. Count the legs on each spider. If you need more legs, drav them in to make 8, then write your number story in the boxes If there are too many legs, cross some out to leave 8 and write your number story in the boxes.

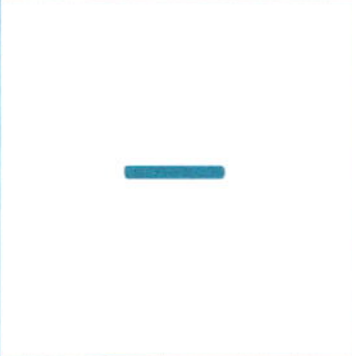

1. Help your child make a spider from plasticine. Use red and blue sticks for legs. How many are red? How many are blue?

Place a sticker here.

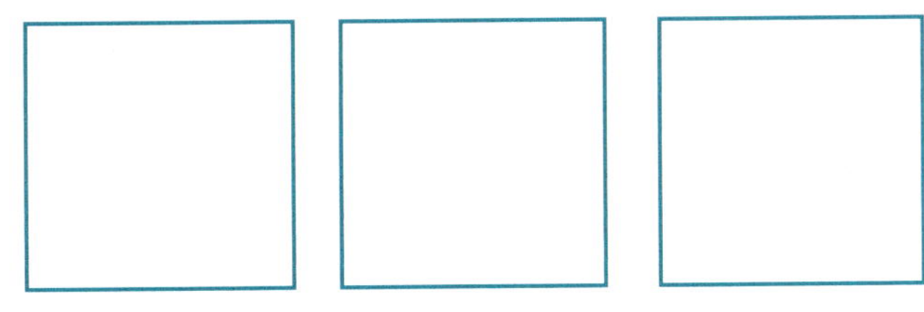

2. How many ways can you show 8 fingers? Help your child find different ways (eg. 5 fingers and 3 fingers).

Place a sticker here.

Making groups of 9

Each tiger should have 9 stripes. Count the stripes on each tiger. If you need more stripes, draw them in to make 9, the write your number story in the boxes. If there are too many stripes, cross some out to leave 9 and write your number story in the boxes.

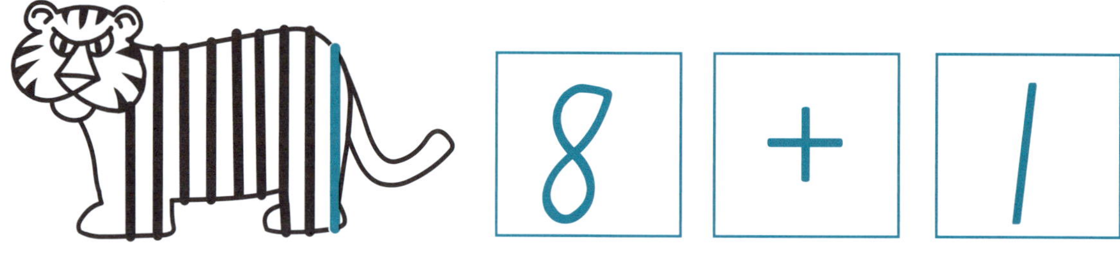

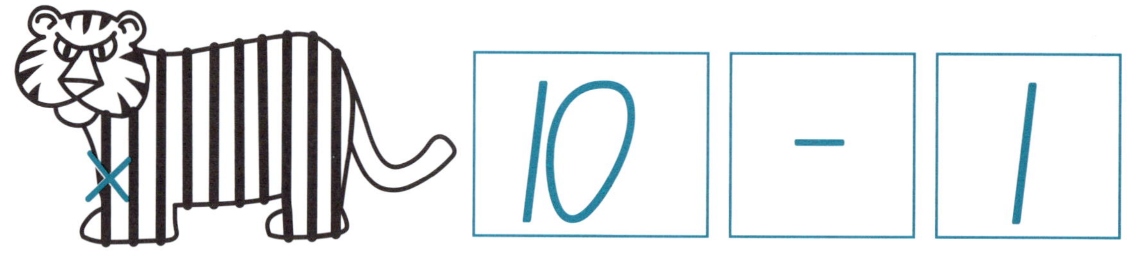

1. Look for pictures or photos of tigers. How many stripes can your child count on them?

Place a sticker here.

2. Draw the outline of a tiger, and help your child paste on sticks for stripes. How many stripes will fit on the tiger?

Place a sticker here.

Making groups of 10

Each cat should have 10 whiskers. Count the whiskers on each cat. If you need more whiskers, draw them in to make 10, then write your number story in the boxes.

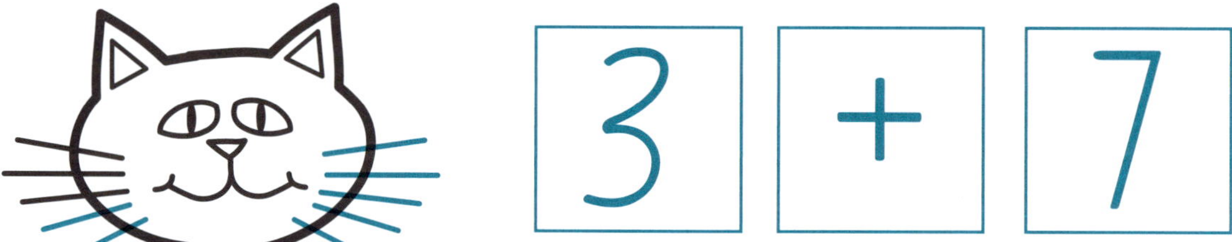

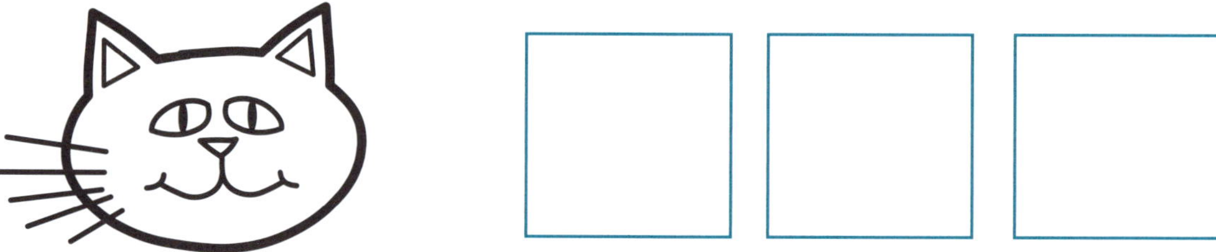

1. Help your child make a cat's face using a paper plate, then paste on sticks for whiskers.

Place a sticker here.

2. Place a collection of 10 objects into 2 groups. Ask your child to count how many in each group, then write a number story for it.

Place a sticker here.

Well done!

You have finished the book!

Place your last two stickers on the picture.
You can colour in the picture, too.